Alan McKirdy has written many popular books and book chapters on geology and related topics and has helped to promote the study of environmental geology in Scotland. His other books with Birlinn include *Set in Stone: The Geology and Landscapes of Scotland* and he is co-author of *Land of Mountain and Flood*, which was nominated for the Saltire Research Book of the Year prize. Before his retirement, he was Head of Knowledge and Information Management at Scottish Natural Heritage. Alan is now a freelance writer and has given many talks on Scottish geology and landscapes at book festivals and other events across the country.

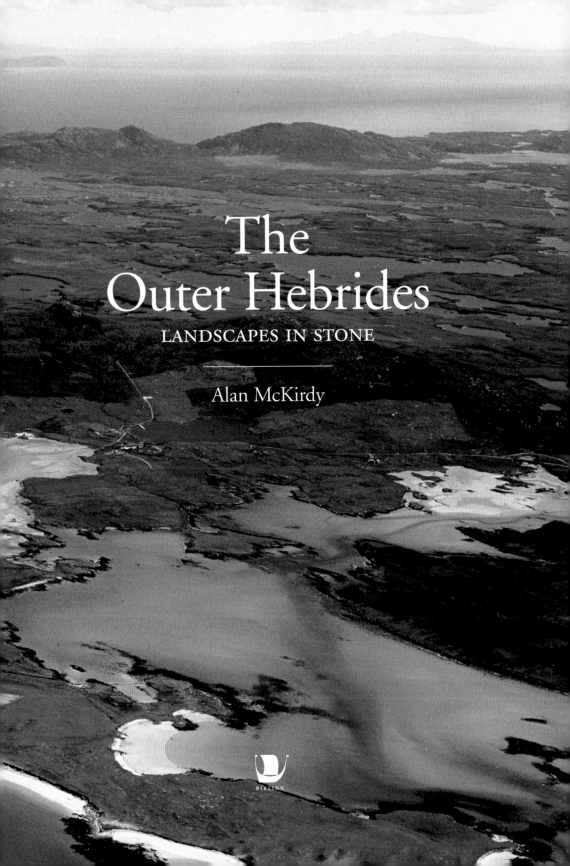

The Outer Hebrides

LANDSCAPES IN STONE

Alan McKirdy

BIRLINN

For Janet Watson

First published in Great Britain in 2018 by
Birlinn Ltd
West Newington House
10 Newington Road
Edinburgh
EH9 1QS

www.birlinn.co.uk

ISBN: 978 1 78027 509 3

British Library Cataloguing-in-Publication Data
A catalogue record for this book is available
on request from the British Library

Designed and typeset by Mark Blackadder

FRONTISPIECE.
Machair of Oronsay,
North Uist

Printed and bound in Britain by Latimer Trend, Plymouth

Contents

Introduction

The Outer Hebrides are a chain of islands of spectacular natural beauty. They stretch from the Butt of Lewis in the north to the small isle of Vatersay, south of Barra, a distance of over 200 kilometres. This archipelago lies off the north-west coast of Scotland and is one of the most remote outposts of the British Isles. The individual islands are of contrasting appearance. Lewis is austere, with a featureless peatland core, bounded by dramatic sea cliffs. Harris is rugged in the extreme, with stunning unspoilt beaches and azure seas. The Uists are characterised by machair lands and sandy beaches running the length of the Atlantic coast. Barra has a more brutal landscape carved from ancient gnarled rock. But these islands have one thing in common. They are all built from the most ancient rocks in Britain – Lewisian gneiss. These are ancient outliers from another world, created when the planet was young and the first primordial crust had just formed around 3 billion years ago. They reach back almost, but not quite, to the beginning of geological time. Their formation took place over an unimaginable timescale – just short of 2 billion years. That's a substantial chunk of the age of the Earth, which was formed around 4.5 billion years ago.

Around 240 million years ago, the land that would become the Outer Hebrides lay just north of the Equator, and layers of desert sands were laid down then that can be seen today to the east of Stornoway.

Evidence of ancient volcanic activity that took place around 60 million years ago is also preserved in the record of the rocks. Rockall, St Kilda and the Shiant Isles are the remnants of ancient volcanoes that erupted when powerful forces ripped continents asunder and created the North Atlantic Ocean.

The Ice Age was a much more recent event, but it had a profound effect on shaping the landscape and creating the dazzling vistas we see today. The characteristic mountains and glens and the softer coastal landscapes owe much to the big freeze and later events.

This book tries to make sense of these ancient events and allows a wider understanding of how these rocks and landscapes came to be.

The Outer Hebrides through time

Period of geological time	Millions of years ago	Scotland's global position	Environments and events in the Outer Hebrides
Anthropocene	Last 10,000	57° N	This is the period when *Homo sapiens* (people), appeared on the scene in the Outer Hebrides and made our presence felt. Such woodlands as existed were felled and a more pastoral way of life was initiated. Thick layers of peat started to accumulate from around 6,000 years ago onwards.
Quaternary	Started 2 million years ago	Present position of 57° N	• **11,500 onwards** – the ice retreated as the climate started to warm. • **12,500 to 11,500 years ago** – the climate became very cold as the ice returned and glaciers once again ground their way across the landscape. • **14,700 to 12,500 years ago** – temperatures were similar to those of today and the ice sheets melted completely. • **29,000 to 14,700 years ago** – this was the last major advance of the ice, with all but the highest peaks of Harris completely covered by ice. The ice sheet extended 70km westwards to the edge of the continental shelf. • **Before 29,000 years ago** – there were prolonged periods when thick sheets of ice completely covered the area. These extreme glacial episodes were punctuated by warmer interludes, known as inter-glacials, when the temperatures rose to levels similar to those of today.
Neogene	2–24	55° N	Conditions were warm and temperate, but temperatures tumbled with the onset of the approaching Ice Age.
Palaeogene	24–65	50° N	Around 60 million years ago, continents split asunder. Volcanoes erupted and spewed lava and ash across the land surface. The bedrock of Rockall, St Kilda and the Shiant Isles was formed during these extreme events.
Cretaceous	65–142	40° N	Sea levels rose to cover the Outer Hebrides, but no rocks of this age are preserved in the Outer Hebrides.

Period of geological time	Millions of years ago	Scotland's global position	Environments and events in the Outer Hebrides
Jurassic	142–205	35° N	Thick layers of sand, mud and limestone were deposited in shallow seas around the Outer Hebrides. The only on-land deposits are preserved on the Shiant Isles.
Triassic	205–248	30° N	Desert conditions prevailed, and layers of conglomerate and associated rocks some 4km thick were deposited by streams and braided rivers.
Permian	248–290	20° N	Desert conditions were widespread.
Carboniferous	290–354	On the Equator	'Scotland' was located at the Equator at this time, but no rocks of this age are preserved here.
Devonian	354–417	10° S	Desert conditions were widespread, but no rocks of this age are preserved here.
Silurian	417–443	15° S	Large upheavals were happening elsewhere across the country as the Highlands of Scotland were formed. Some late movement on the Outer Hebrides Fault took place at this time.
Ordovician	443–495	20° S	'Scotland' was located on the southern shores of the Iapetus Ocean, which had started to close by this period.
Cambrian	495–545	30° S	No rocks of this age are preserved in the Outer Hebrides.
Proterozoic	545–2,500	Close to South Pole	The Lewisian gneisses formed during Archaean times were reworked by later mountain-building events. Granites of the Uig Hills were added to the bedrock during these times. Movement was initiated on the Outer Hebrides Fault.
Archaean	Prior to 2,500	Unknown	The Lewisian gneisses built most of the islands of the Outer Hebrides. These rocks were formed, then folded and faulted over a period of just short of 2 billion years. The oldest rocks in the Lewisian complex were formed around 3 billion years ago. The age of the Earth is around 4,540 million years.

Geological map of the Outer Hebrides. The Outer Hebrides are largely built from ancient rocks – the Lewisian gneisses. These rocks were forged deep within the Earth's crust and brought to the surface by later contortions within the crust and subsequent erosion by ice, wind and water. They were mostly formed from even more ancient molten rocks and later altered by high temperatures and pressures during their prolonged burial at considerable depths below the surface. The gneisses largely consist of light and dark bands, interspersed with veins of brightly coloured pink granite and darker, more base-rich, bands of dolerite. South and west Harris are built from granite, which forms much of the spectacular high ground. A variety of granite known as anorthosite is found near Rodel on Harris. Unweathered, it is ice-white and a dramatic addition to the landscape of the area. Later, as 'Scotland' drifted northwards, desert conditions prevailed and rivers ran across the parched land. They deposited thin layers of sediments that built the Eye peninsula. Then, as sea levels rose, Jurassic sediments were deposited in the Minch between the mainland and the Outer Hebrides. A small outpost of this rock occurs on the Shiant Isles but it's too small to show up on the geological map. Rockall, St Kilda and the Shiants were formed as volcanoes erupted and rocked the area around 55 million years ago.

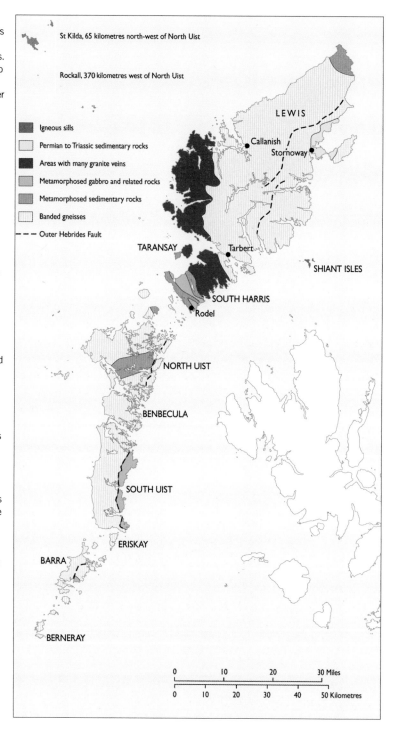

St Kilda, 65 kilometres north-west of North Uist

Rockall, 370 kilometres west of North Uist

- Igneous sills
- Permian to Triassic sedimentary rocks
- Areas with many granite veins
- Metamorphosed gabbro and related rocks
- Metamorphosed sedimentary rocks
- Banded gneisses
- – – – Outer Hebrides Fault

LEWIS

Callanish
Stornoway

TARANSAY   Tarbert

SHIANT ISLES

SOUTH HARRIS
Rodel

NORTH UIST

BENBECULA

SOUTH UIST

ERISKAY
BARRA

BERNERAY

0   10   20   30 Miles

0   10   20   30   40   50 Kilometres

1
Time and motion

Time

The structure of standing stones at Callanish (Calanais) on the Isle of Lewis is one of Scotland's most iconic monuments. It was built around 5,000 years ago and consists of a stone circle with a long avenue of erect stones. It is thought to have been the focus for early religious activity for over 1,000 years.

Although it may seem that our human history represents the edge of time, study of the rocks and landscapes can extend that timescale millions or even billions of years further back. The earliest events in the geological history of Scotland are recorded in the rocks of the Western Isles. The rocks of Lewis and Harris have been dated at more than 3,000,000,000 (3 billion) years old. It's difficult to fully understand these numbers when we see historical events through the prism of human timescales. But learning to appreciate this extended timescale is an essential part of studying geology.

This stone circle and associated avenue of standing stones at Callanish on Lewis was built from local stone – Lewisian gneiss. This structure is considered to be unique as it has been constructed in the shape of a Celtic cross. Thirteen stones, between 3 and 4 metres in height, form a circle with a diameter of 13 metres. The erection of impressive structures such as this would have underlined the power of local leaders and enhanced the cohesion of the community.

The peak at Roineabhal is perhaps not typical of the scenery of the Outer Hebrides. It is scarred by quarrying and the creation of access roads, but apart from these limited activities, it is largely untouched since the ice melted. The summit was moulded and shaped by the passage of ice that melted around 11,500 years ago.

Geologists throw around dates of 'millions and billions of years ago' that are unfamiliar and perhaps even incomprehensible to most people. But they are now an accepted part of the science of geology. Arthur Holmes, Regius Professor of Geology at Edinburgh University during the 1950s and 1960s, was at the forefront of the development of techniques to date rocks accurately. This ground-breaking work enabled a better understanding of the way in which the sequence of geological events unfolded over time. The rocks of the Outer Hebrides were formed over a very protracted period of time. Being able to date rocks and related events accurately, or within a window of plus or minus a few million years, allowed us to understand the geological history of the Outer Hebrides and to place events in a logical chrono-logical order.

Motion

The Earth's crust is made up of a series of individual slabs of rock known as tectonic plates. Ours is a dynamic world where these plates are driven across the surface of the planet by the motion of the underlying mantle. This has the effect of constantly re-arranging the geography of the planet as these plates glide and jostle their way across the globe. This theory of how the Earth works is known as plate tectonics.

Above. The Earth's surface is divided into a series of tectonic plates – seven large ones and a dozen or so smaller ones. Driven by the heat engine of the Earth's core and mantle, these plates grind past each other, creating daily minor earthquakes and larger periodic catastrophes such as tsunamis and larger, more destructive earthquake events. These movements push the individual plates across the surface of the Earth at an average rate of around 6cm per year. Over billions of years, the land that was to become the Outer Hebrides moved from a position close to the South Pole through every climate zone the Earth has to offer, to its present location 57° N of the Equator.

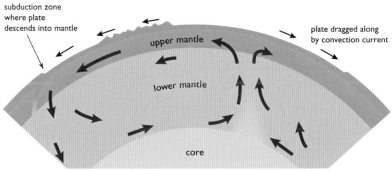

Left. The Earth's tectonic plates are moved across the globe, driven by heat that comes from the centre of the planet. The mantle actively moves in response to this heat flow from the Earth's core, and the plates are dragged along as a consequence. Over geological time, continents have moved many thousands of kilometres as a result of this process.

As continents collide, the sediments that accumulated on the ocean floor are folded and buckled to form a mountain chain. The Highlands of Scotland were formed in this way and were as impressive as the Himalayas are today when they were first formed around 420 million years ago.

On their journeys, continents collide with each other. The thick accumulations of sands, muds and limestones that built up on the sea floor over many millions of years were crumpled, folded and faulted as the continents met head-on. The mountains of the world provide evidence of this process having been repeated throughout geological history. The Himalayas, for example, were formed as a result of a continent-to-continent collision of colossal proportions when India split from the side of Africa and moved northwards to bulldoze into Asia. The resultant mountain chains are described as 'fold mountains' as the layers of soft sediments were folded, cooked and squashed in this extreme environment.

As we explore the geology and landscape development of the Outer Hebrides, it's important to bear in mind these important factors of time and motion. Scotland has travelled the globe during its 3 billion year history and where it is now, 57° N of the Equator and subject to, at best, changeable weather, is just one point on its journey. In millions of years to come, its position on the globe will be very different from its current location, as will the climate and general environment. That's just the way our dynamic Earth works.

2
Ancient worlds

The ancient crust that was to become the Outer Hebrides dates back to what geologists called Archaean times. The Earth had been bombarded by comets and asteroids for many millions of years as the early processes that created the planet played out. But, around 3 billion years ago, stable continents became established and with that development came a greater sense of order. The surface of the Earth was now divided into land and sea.

The Archaean oceans were more extensive than the oceans of today and the area covered by dry land was commensurately smaller. The world's oceans were twice as salty as today and entirely devoid of oxygen. Any life that existed in these oceans would have been most primitive in nature, probably single-celled bacteria that did not require oxygen to survive. Volcanic vents, known as black smokers, billowed columns of super-heated water from the sea floor that were laden with iron sulphides and other toxic chemicals.

Active volcanoes pock-marked the early crust in the Archaean world. The resultant atmosphere lacked oxygen and it took many millions of years before the planet was able to support life on land and in the seas.

Hydrothermal vents or black smokers pumped super-heated water loaded with sulphides and other toxic chemicals into the early oceans.

The atmosphere, too, would have been devoid of oxygen, consisting largely of nitrogen, with some methane and carbon dioxide. The Sun was much weaker then, so without an insulating layer of greenhouse gases in the atmosphere the Earth would have remained a frozen ball of rock. However, in this greenhouse world, temperatures rose to the dizzy heights of up to 80°C. Then, as oxygen started to be produced by photosynthesising bacteria, the Earth lost heat and the temperature plummeted to a point where the Earth's first global glaciation took place. It was in this confused and deeply hostile world that the bedrock of the Outer Hebrides was forged.

Determining with any great accuracy the location of the continents that existed worldwide in Archaean times is difficult, but not impossible. Studies of rocks from around the world has allowed geologists to identify the earliest areas of dry land and understand how they may have once been related to one another, forming coherent slabs of continental crust. As we've seen in the previous section, rocks can be dated with a high degree of accuracy, so a jigsaw of sorts can be assembled representing the continents that existed during these far-off times. This tiny corner of Scotland sat alongside Greenland, the Canadian Shield and areas of Scandinavia as survivors of the turbulent Archaean period.

In terms of the overall development of the land that was to become Scotland, this was the basement – the very lowest layer. It doesn't get

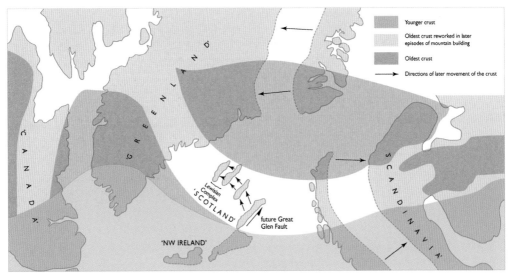

This reconstruction of ancient continental fragments shows the relationship between the 'land that was to become Scotland' with chunks of crust of similar age in Canada, Greenland and Scandinavia. All are survivors from the earliest turbulent events to have been recorded on Planet Earth and each segment of crust has been squeezed and squashed (metamorphosed) many times. The imprint of each event is 'remembered' in the alignment of minerals (called foliations) and the formation of specific mineral associations and combinations. These indicators allow geologists to calculate the age of individual rocks sampled and the depth at which they were formed in the Earth's crust. All these ancient fragments of crust were located close to the South Pole!

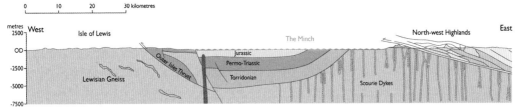

This slice through the Earth's crust runs from west to east traversing the Outer Hebrides and running across the Minch to the mainland of Scotland. The term 'basement' applied to the Lewisian gneiss is most apposite as these are the lowest and oldest layers of crust to be found anywhere in the British Isles. These basement rocks of Lewisian type are thought to extend as far south and east as the Great Glen. Younger rocks were piled on top. The Minch is an expanse of water between the Outer Hebrides and the mainland. It is floored by layers of sediment from younger times – the Torridonian, Permo-Triassic and Jurassic. How these layers were formed and their part in the story will be told in later sections.

any older than this. If we think of prominent landmarks from around Scotland, such as Ben Nevis, the Cairngorms, the Old Man of Hoy or Staffa, many millions of years would elapse before any of these features came into being. So these early events formed the foundation layer on which all later events were played out.

Historical study of the rocks

The Lewisian gneisses of the Outer Hebrides have been intensively studied in modern times, but things got off to a slow start. J. MacCulloch was the first to publish a basic account of the geology in 1819. Thereafter, these outposts were largely ignored until 1923 when a series of detailed accounts, prepared by T.J. Jehu and R.M. Craig from the University of Edinburgh, were published in the journal *Transactions of the Royal Society of Edinburgh*. Academics from Imperial College in London were next on the scene, in conjunction with the Institute of Geological Sciences (now the British Geological Survey). A comprehensive set of reports and geological maps was generated as a result of this collaboration. It is the combination of these studies, old and new, that has allowed us to piece together the turbulent geological past of the Outer Hebrides. This book is dedicated to one of those pioneers – Professor Janet Watson. In the 1940s and 1950s, she made her name and reputation in unravelling the complexities of these ancient rocks, both on the Outer Hebrides and on the Scottish mainland. She died aged 61 at the peak of her powers, but left behind a formidable canon of work.

Lewisian bedrock

The first events in the Lewisian story are dated at around 3 billion years ago, which is an extraordinary date to conjure with. The Earth is considered to be some 4.54 billion years old, so the earliest events recorded in the Outer Hebrides reach back almost to the beginnings of the planet itself. The oldest rocks are light-coloured granites and darker base-rich rocks (rocks made from minerals rich in calcium, magnesium and iron) that were added to the lowest layers of the crust around 3 billion years ago. In fact, by far the greater part of the basement was originally molten material. Some parts, however, started as sediments, such as limestones and layers of mud.

These original attributes were significantly modified by deep burial in the Earth's crust forming characteristically banded rocks of alternating dark and lighter layers. The paler layers are made from crystals of quartz and feldspar, two of the commonest rock-forming minerals, with the darker bands made largely from a mineral known as amphibole. These layers, or foliations as they are known, formed at right angles to the direction in which they were squashed at depth in the Earth's crust.

Top. This close-up of the layering or foliation shows the alternation of light and dark bands that are typical of the Lewisian gneiss.

Middle. The banded nature of the Lewisian gneisses is beautifully illustrated by these rocks at Rubh' Aird-mhicheil on South Uist.

Bottom. Pulses of base-rich magma were introduced to the basement rocks around 2.4 billion years ago and have proved to be invaluable in unravelling the complex geology of the area. In some locations, these vertical sheets of magma, known as Scourie dykes, have been altered by later metamorphism, but in other places they are unaffected by later events. So events that happened before the dykes were introduced into the bedrock can be distinguished from later episodes of folding and faulting. It's a very simple way of splitting events into 'pre' and 'post' the intrusion of the accurately dated Scourie dykes. This observation revolutionised our understanding of the sequence of geological events that gave rise to these rocks. It was Professor Janet Watson, whose contribution was described earlier, who first realised the value of the Scourie dykes as a potential time marker.

The light and darker coloured rocks have been bent double by events that took place deep within the Earth's crust. This cliff section, which is located on the north-west coast of North Uist, is around 30 metres high and demonstrates the powerful forces at work.

Walking on the moon!

Roineabhal is the most prominent hill on South Harris, rising to some 460m above the sea. Roineabhal is composed mainly of anorthosite – an uncommon variety of granite that is rich in the ice-white mineral feldspar. This once-molten intrusion was added to the bedrock around 1.8 billion years ago and later bent double into a tight fold. Rocks of very similar geology to that of Roineabhal make up the highlands of the Moon, so there is something that is truly 'out of this world' about this place!

Opposite top. Roineabhal is a major feature of the South Harris landscape. It towers above the surrounding landscape, scraped clean by the passage of ice and largely unencumbered by vegetation of any sort – a geologists' paradise!

Opposite bottom. This view of the west coast of Harris shows the nature of the sheets of granite that make up this complex of rocks. Since they were added to the bedrock some 1.6 billion years ago, erosion has removed the overlying burden to reveal them at the surface. In much more recent times, the high ground has been scoured by ice and, as seen in this picture, the sea has carved these ancient rocks into a series of skerries and stacks.

This is unspoilt land, bearing only a few minor surface scratches from past episodes of rock extraction. But this tranquillity was threatened in the 1990s by a proposal to create Scotland's second superquarry. Ready access to deep water and international shipping lanes made this an ideal location to quarry road stone on an industrial scale for export to Europe and North America. However, the application was refused, and this place, with its lunar connections, lives on untouched by our clumsy and inappropriate interventions.

Granites of the Uig Hills

The mountains and hills of Harris are hewn from more recent rocks, but only comparatively so as they date from some 1.6 billion years ago. A series of granite intrusions extend over an area of 420 square kilo-

metres, from the Uig Hills of Lewis southwards including the island of Scarp to South Harris. These pillar-box-red granites are magnificently exposed in the glacially scoured hills and dramatic coastline of the area. Countless sheets of granite, varying in size from a few centimetres to hundreds of metres in thickness, were added to the bedrock in a frenzy of volcanic activity. These events happened at depth within the Earth's crust, so there were no erupting lavas at the surface to mark what was happening many kilometres below. There were clearly many episodes of granite intrusion as the sheets are of differing mineral composition and grain size.

The ground moved

One of the final events that formed the bedrock of the Outer Hebrides was the initiation of a major crack in the Earth's crust that runs the

When the Outer Hebrides Fault moved, the temperatures generated along the line of the fault plane were so intense that the rock melted. The darker areas on this rock face are patches of rock glass created by that process. This unusual rock is called pseudotachylite.

Eaval on North Uist shows the line of the Outer Hebrides Fault. The lower surface of the fault line was toughened by the formation of melted rock as the fault moved. As a result, this lower surface of the fault zone was more resistant to erosion when the area was covered and eroded by ice in much later times. The eastward-facing low-angled slope of Eaval preserves the plane of the Outer Hebrides Fault in dramatic fashion. It is most unusual to see the trace of a fault as a prominent feature of the landscape.

length of the island chain. We recognise it today as the Outer Hebrides Fault. The trace of this landscape feature is shown on the geological map. This weakness in the rocks was (possibly) initiated around 1.1 billion years ago and it has moved periodically ever since. As the rocks ground against each other on either side of the fault, the heat generated by this process was so great that there was localised melting of the rocks along the points of contact. This created a very unusual rock known as pseudotachylite. The melted rock cooled quickly to form a network of glassy veins. The forces in the Earth's crust that initiated this movement have now dissipated, so it is most unlikely that there will be further disturbance along this fault line.

This melting of the rock made the surfaces tougher and more resistant to erosion. A line of low hills now marks the location of the Outer Hebrides Fault from Barra to North Uist.

3

Desert sands, then rising seas

The Lewisian gneiss complex was formed over a period of 2 billion years from the earliest datable event to the final occurrences that shaped these basement rocks. A huge chasm of time opened up before the next events made their mark on the Hebridean landscape. By Triassic times, some 248 million years ago, Scotland had migrated north of the Equator, driven on relentlessly by the conveyor-belt movements in the mantle, the layer below the crust. The Lewisian gneisses now formed a small part of a much larger continent that was an amalgamation of all the continents across the face of the Earth. It was dry land from pole to pole in a landmass known as Pangaea, or *All Earth*, from the ancient Greek.

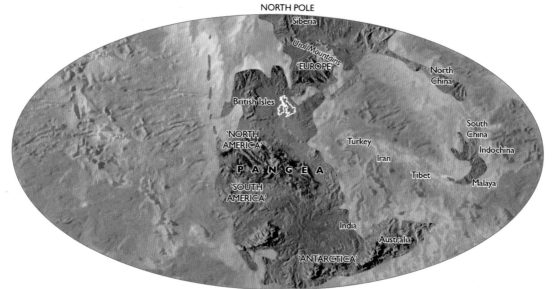

During Triassic times, some 248 million years ago, Scotland had drifted northwards to a location around 30°N of the Equator. In these latitudes, desert conditions prevailed. The land was parched, but seasonal floods were also known to have occurred. Rivers that built the Triassic deposits ran from the west across an area that is now Greenland and North America.

During Triassic times, global warming produced one of the hottest periods in Earth's history with average temperatures approaching 40°C in the area where the land that was to become Scotland was located. The Eye peninsula, just to the east of Stornoway, is built from rocks that were deposited at this time. Seasonal rivers criss-crossed this parched land, dumping deposits of rounded pebbles and layers of sands in the form of a series of debris fans. High ground lay to the

These thick accumulations of conglomerates are found near Stornoway. They were deposited during Triassic times when seasonal rivers ran across the desert landscape.

Cracks appeared as Pangaea began to break up. Sea levels rose dramatically at this time and the Minch was established as a new inlet of the sea. This was the initiation of a new seaway that was eventually to become the North Atlantic Ocean.

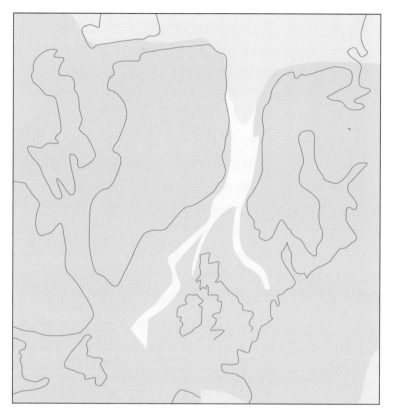

west, and a river carrying the debris tumbled over a cliff edge created by a fault-line in the bedrock. Piles of pebbles, cobbles and larger lumps of rock up to almost a metre in diameter accumulated in a series of fans that built out from the cliff edge. Thin fossil soils, perhaps containing roots, have also been discovered in these deposits, indicating a period of greater stability where a covering of vegetation may have briefly flourished. The cross-section through the crust on page 17 also shows that a considerable thickness of Triassic sediments was dumped in the area now occupied by the Minch. So this period of sediment accumulation must have persisted for many millions of years.

As the Triassic Period drew to a close, sea levels began to rise substantially. During the Jurassic Period that followed, there was a dramatic re-arrangement of the continents as Pangaea began to break up and great cracks appeared, splitting the continent asunder.

As sea levels rose, water flooded the area now occupied by the Minch and formed a shallow sea. Layers of mud, sands and limestone were dumped on the desert sediments of earlier times. Jurassic sediments have been recovered in drill cores from the sea bed of the Minch and are also seen on neighbouring Skye. But on-land occurrences of these rocks in the Outer Hebrides are limited to the Shiant Isles, 20km east from Harris. The ancient Jurassic seas teemed with life. A variety of corals, oysters, ammonites and bullet-shaped belemnites are preserved as fossils. These Jurassic deposits connect with the oil-bearing strata that are currently being explored and exploited west of Shetland.

Jurassic rocks also accumulated to the west of the Outer Hebrides in a trough called the Rockall Rift. As the continental re-arrangement continued and Pangaea cracked open further, channels of the sea extended further inland. But the Outer Hebrides remained above the waves, as part of a wide platform of dry land that extended some 200km to the west of the island chain.

The rise of sea level continued and indeed accelerated into the Cretaceous Period. This was a greenhouse world where the ice caps melted and sea levels rose to an astonishing 200m higher than they are today. When no evidence exists, it becomes a matter of speculation, but it seems likely that the Outer Hebrides would have been completely covered by a warm shallow sea for a few million years or so. No deposits of this age have been recorded anywhere on the archipelago. What is much more certain, however, is what happened next.

This ammonite lived in the seas just to the east of the Outer Hebrides. Its remains are held in a matrix of mud, now squashed to form shale.

4
Ancient volcanoes

The continental re-arrangements continued as Pangaea continued to fracture and new oceans opened. For many millions of years, 'North America' and 'Europe' had been locked in an embrace, but now a widening ocean intervened. The driving force for this process was once again to be found deep within the Earth. Rising streams of super-heated rock ascended through the Earth's mantle and reached the surface in a frenzy of magma and superheated seawater. This phenomenon is known as a mantle plume. Strips of basalt lava were added to the crust below the ocean as the upwelling of molten rock from the bowels of the Earth continued. And with each new pulse, the continents inched further and further apart. Many active volcanoes appeared in the north-west corner of Scotland, including Skye, Mull,

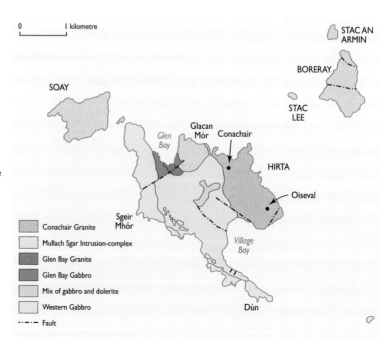

This geology map of St Kilda shows the compositional variations that evolved in the magma chamber that sat below the volcano. The various granites and gabbros developed at slightly different times during the lifetime of the volcano. The last to form was the Conachair granite that makes up the eastern part of the island. Soay and the western parts of Hirta, the main island of St Kilda, are made from base-rich gabbros and related rocks.

0   I kilometre

STAC AN ARMIN

BORERAY

SOAY

STAC LEE

Glacan Mór   Conachair

Glen Bay

HIRTA

Oiseval

Sgeir Mhór

Village Bay

Conachair Granite

Mullach Sgar Intrusion-complex

Glen Bay Granite

Glen Bay Gabbro

Mix of gabbro and dolerite

Western Gabbro

--·—· Fault

Dùn

Ardnamurchan and Arran. Rockall and St Kilda were also active volcanoes at this time. This volcanic episode started around 61 million
years ago and lasted at the various sites for up to 6 million years.

St Kilda was one of the last to blow its top. Its peak activity is
dated at around 55 million years and it was hot for around another
million years. Volcanoes are fed by large underground storage areas
for molten rock, called magma chambers. This boiling cauldron cooled
slowly at depth and a variety of rock types solidified from this melt,
reflecting the changing composition of the magma at different times
during the evolution of the volcano. Granite and its more base-rich
and blacker equivalent gabbro were the most common products of
this process. There are no known lavas associated with the St Kilda
volcano. Only the magma chamber survives today. Lavas may have
been erupted, only to be swept away by later erosion.

Some of this molten rock didn't quite reach the surface and was
squirted sideways to form a layer inserted into the Jurassic sediments
that were already in place. These pulses of magma cooled slowly and

Sea cliffs have been cut by
ice, wind and water into the
ancient St Kildan volcano.
Today, these dramatic
features provide a refuge
and nesting areas for the
many sea birds that inhabit
this bleak outpost in the
midst of the stormy North
Atlantic Ocean.

View of the layers of igneous rock that were inter-leaved between sediments of Jurassic age on the Shiant Isles.

solidified to form what are known as sills. The Shiant Isles are composed almost exclusively of igneous rocks formed in this way. The thickest sill is around 150m wide and is well displayed on steep cliffs on the largest island of Garbh Eilean.

Rockall lies around 300km west of St Kilda and is the furthest outpost of the UK. It, too, was an ancient volcano from this era, which has been eroded down to an isolated granite stump that projects a mere 20m above the Atlantic swells. At 30m across, it is a much more modest volcanic edifice than its siblings of Skye and Ardnamurchan. But it was formerly a grander affair as related volcanic rocks have been found on the adjacent sea bed, which were at one time an integral part of this ancient volcano. The open ocean between Rockall and the mainland is floored by sedimentary rocks of Jurassic age. The area has been investigated as a potential oil reservoir, although no commercial oil extraction has taken place to date.

Left. Some sills on the Shiant Isles exhibit spectacular columns that look like a bank of organ pipes carved into the rock. They are similar structures to those found at Fingal's Cave on Staffa and the Giant's Causeway on the Antrim coast. These features form when molten magma cools slowly, allowing natural forces to organise the rock into hexagonal structures.

Below. Rockall is located around 300 kilometres west of the Outer Hebrides. It is around 20 metres high and 30 metres across. Like St Kilda, Skye, Mull, Ardnamurchan and Arran, this place was an active volcano around 55 million years ago, belching ash and volcanic bombs high into the air. Since that time, erosion has pared this giant down to size and all that is left is a small stump of granite that originally formed part of the magma chamber of the Rockall volcano. The seafloor around Rockall and between there and the Scottish mainland is punctuated by a dozen or more smaller igneous centres (ancient volcanoes) that were active around the same time as Rockall but aren't of sufficient height to break the surface of the ocean. These structures are described as 'seamounts'.

5

The Ice Age

After the temperate climate and active volcanoes of the Palaeogene Period, temperatures declined gradually as the Ice Age took hold. Over a period of around 2.6 million years, Arctic conditions prevailed through much of continental Europe. The climate fluctuated between lengthy cold periods when the land was under thick layers of ice and snow throughout the year and periods when temperatures were more similar to those of today. Much of the water on the planet was locked up in glacial ice and snow, causing global sea levels to fall.

These dramatic changes in temperature and climate are explained by the Earth's eccentric orbit around the Sun, which is sometimes perfectly circular, and at other times elliptical. The transition from circular to elliptical orbit takes around 100,000 years. The temperature on the Earth's surface is clearly influenced by its distance from the Sun. When the Earth is further from the Sun, temperatures fall, sending the Earth into a glacial phase where ice sheets spread at the poles and ice and snow accumulate at lower latitudes. Variations in the Earth's tilt and the wobble of its axis of rotation also affect the amount of solar radiation that falls on the surface of the planet. There have been many periods when the Earth has been gripped by glacial conditions, each separated from the previous advance of the ice by a period of relatively warmer conditions, known as an interglacial period. The Earth is currently in an interglacial phase, but glacial conditions are expected to return at some time in the future. Current estimates are that the next advance of the ice, which will engulf Britain from north to south, will begin in around 50,000 years' time.

Around 22,000 years ago, the last glaciation reached its peak in

Opposite. This is Antarctica today, but it also shows how Scotland might have looked during the last advance of the ice. These advances and retreats of the ice had a significant effect in shaping the landscape, with mountains and glens carved and moulded by the passage of the ice. The highest hills on Harris and South Uist would have poked through the cover of ice and snow in the manner shown in this image.

The ice scraped across the landscape of South Harris, shaping the bedrock into the hills and glens we see today. In most other parts of the country, the erosive effects of the ice have been masked by later man-made modifications to the landscape, such as farming, forestry and development of infrastructure such as roads and settlements.

this area when vast ice sheets flowed westwards from the highest ground towards the sea.

After the ice

Abruptly, around 14,700 years ago, temperatures rose, melting the world's ice caps. Summer temperatures were similar to those of today. As the ice melted, boulders, stones and mud carried by the ice were dumped forming characteristic piles of glacial debris, known as moraines.

After another brief period of lower temperatures, the ice melted in this area some 11,500 years ago. A key factor in this slow warming process was a flood of warmer waters emanating from the Gulf of Mexico, which flow across the Atlantic Ocean. This ocean current is known as the Gulf Stream. We feel the benefits of its benign influence to this day.

Pollen grains and insect remains recovered from the peat bogs across the islands has allowed scientists to chart the changing conditions that existed during these turbulent times. First to colonise the chaotic 'moonscape' of stony deposits dumped by the ice were herbaceous plants followed by heather, juniper and grass. Then came birch, and later hazel and oak. By 8,000 years ago, forest covered most of the islands. The activities of the species *Homo sapiens* (us!) during Mesolithic times reduced the forest cover substantially. Around 6,000 years ago, the climate became cooler and wetter, encouraging an expansion of moorland and the development of a blanket of peat bog. The tree cover had probably disappeared almost completely by 3,000 years ago as the development of blanket bog continued apace.

The water liberated as the ice melted also helped to shape the landscape. Perhaps the best example is found on the road to Uig at Glen Valtos.

Tree stumps found in peat are clear evidence of a once widespread cover of woodland.

Above. These flat plains would have been unrecognisable 8,000 years ago. Birch, hazel and oak would have covered the landscape.

Right. This meltwater channel at Glen Valtos was carved into the bedrock by water that came from the melting ice.

6
Landscapes today

Cultural landscapes

St Kilda was inhabited from the very earliest times. An earth-house was excavated in Village Bay that dates from 5000 BC. After that date, there was a changing population of permanent inhabitants until the dwindling numbers of people living on Hirta were evacuated in August 1930. The islands were bequeathed to the National Trust for Scotland in 1957 and received World Heritage status in 1987. In terms of the geology, the northern side of the bay (left-hand side of the picture) is composed of pale pink to white granite and the ground rises steeply to the summit of Conachair.

Callanish has been called the 'Stonehenge of the Hebrides'. It is a

This sweeping panoramic shot of Village Bay on St Kilda demonstrates the human imprint that has been left on this place. The black houses, cleits (built as storage facilities), dry-stane walls and other structures have been the subject of archaeological excavations over many decades, and that work continues to this day.

The purpose and function of these standing stones is still an enigma, but there seems little doubt that the Sun, Moon and the stars had some role in whatever dramas unfolded here during Neolithic times.

The contortions in the banding of the Lewisian gneiss are clearly visible in the flank of this standing stone at Callanish.

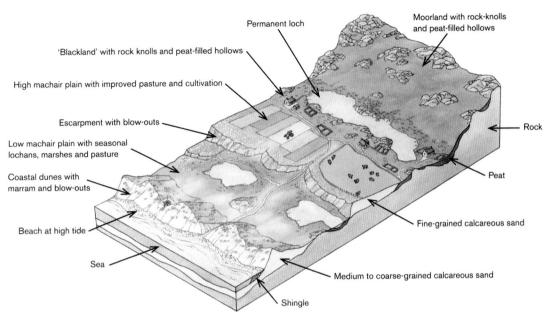

Permanent loch

Moorland with rock-knolls and peat-filled hollows

'Blackland' with rock knolls and peat-filled hollows

High machair plain with improved pasture and cultivation

Escarpment with blow-outs

Low machair plain with seasonal lochans, marshes and pasture

Coastal dunes with marram and blow-outs

Beach at high tide

Sea

Rock

Peat

Fine-grained calcareous sand

Medium to coarse-grained calcareous sand

Shingle

Above. This diagram of the machair is a good representation of the main elements of the landscape – from the sea to the rocky knolls inland.

Below. The machair in full bloom at Stilligarry on South Uist.

Aerial view of the machair landscape at Sollas and Malaclete, North Uist.

hugely significant cultural landscape where geology has played a key role in construction of this ancient monument. The standing stones were erected here about 5,000 years ago, consisting of a central monolith, surrounded by 13 stones and a series of avenues that radiate from the central point. The stones are all of Lewisian gneiss, planed in the direction of the natural banding, known as foliations. This monument is an incredible survivor from the 'megalithic' culture with structures of a similar age at the Ring of Brodgar in Orkney, Castlerigg in the Lake District and of course Stonehenge itself.

The machair of the Outer Hebrides is a unique form of cultural landscape that has existed here for centuries. The soils were formed after the ice melted, from broken shell debris transported onshore by wind and waves. The fertile land of the machair is also a working landscape, supporting the livelihoods of many crofters who eke a living from the limy soil. The species-rich grasslands are farmed in a very low-intensity manner, in some cases by many generations of the same family. Fertility of the soil has been improved by the addition of seaweed, and occasional drainage works facilitate the cultivation of every last parcel of land. Machair grasslands extend the length of Lewis and Harris, albeit not in a continuous carpet, but as little pockets where conditions allow. In the Uists, however, the machair habitat runs in almost unbroken fashion along the Atlantic coast.

Dramatic seascapes

The four islands of St Kilda – Hirta, Dun, Soay and Boreray – create some of the most dramatic seascapes in the UK. Conachair on Hirta is some 430m high and a sheer sea cliff on its northern flank falls vertically to the sea below. Stacks, arches, geos, tunnels, caves and blowholes add considerably to the complexity of this dramatic seascape. The outline of these islands was marked out by glacial activities during the Ice Age, but it has been modified every day since by the incessant pounding of the Atlantic waves and howling wind in this most exposed of locations.

The seacliffs of the islands of St Kilda teem with seabirds. Gannet, fulmar, guillemot, Leach's petrel, puffin, razorbills and storm petrel all make these cliffs their home base. They are significant in terms of the overall breeding populations of many of these species of seabird. The cliffs also support a thriving population of plants with a coastal distribution such as thrift, sea campion and ribwort plantain.

Natural landscapes

The core of the Isle of Lewis is covered by a thick blanket of peat. This habitat is very rare internationally and this featureless stretch of moorland represents a significant proportion of the world's peat resource.

The blanket bog is punctuated by areas of open water which are home to a range of rare and endangered species of birds. Red-throated divers can be found here and the rare black-throated divers can even be spotted catching fish in the lochans. The iconic golden eagle also has a toe-hold, as have the re-introduced white-tailed eagles, also

The golden eagle is a magnificent sight. Languid wing beats and frequent glides exploiting up-draughts of air are characteristic of their pattern of flight. Individuals can have a wingspan of over two metres, with dark brown plumage, a powerful beak and large talons.

Red-throated divers nest in freshwater pools but feed at sea. During the winter season, they are largely marine birds. Their distribution is limited to northern Scotland and the islands, although they are more widespread in Iceland and throughout Scandinavia.

known as sea eagles. Infrastructural developments have not yet cluttered this place, although wind-energy companies have eyed up these broad empty spaces with a degree of longing.

The peatland habitat, which is also widespread in other parts of the Outer Hebrides, provides fuel for the fires of many homes in these islands. The peats are cut from the moor, normally by hand, and stacked to dry in the frequent storm-force westerly winds.

The corncrake is an iconic species of this place. Active conservation measures have helped to increase numbers of breeding pairs to a much greater level than previously. Reduction of the predator popu-

Corncrakes are widespread throughout the areas of machair on the Atlantic coast of the Outer Hebrides. This low-intensity farmed grassland provides the ideal habitat to support this species. Many people who live in their midst may never have seen these secretive birds but may have heard their distinctive call. Birds that breed here migrate to the grasslands of Africa, leaving in September and returning in April or May.

Stacked peat freshly cut from the bog in North Uist.

lation, particularly mink, has helped considerably in creating a sustainable population of these birds. The corncrake's scientific name is *Crex crex*, which some say is redolent of the call that the males make during the breeding season.

Literary landscapes

The rocky island of Eriskay, just to the north of Barra, is the setting for *Whisky Galore,* one of Scotland's best-loved tall tales. There is, however, more than a grain of truth in the story. During the Second World War, the SS *Politician* was wrecked on the rocks off the coast of Eriskay with a cargo of 50,000 cases of whisky aboard. The author Compton MacKenzie, who lived on neighbouring Barra, embroidered the tale to become an epic of wily locals pitting their wits against the authorities. The Home Guard Captain Paul Waggett tried to retain order while the locals grabbed and consumed as much of the cargo as they possibly could.

The reefs offshore from the island of Eriskay were the undoing of the SS *Politician*, which crashed onto the rocks more than 70 years ago.

7
Places to visit

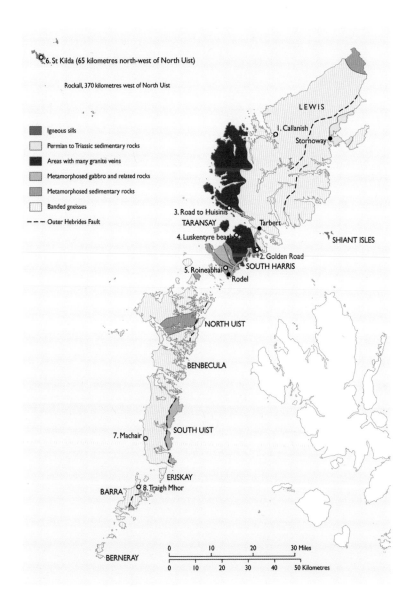

6. St Kilda (65 kilometres north-west of North Uist)

Rockall, 370 kilometres west of North Uist

LEWIS

1. Callanish
Stornoway

- Igneous sills
- Permian to Triassic sedimentary rocks
- Areas with many granite veins
- Metamorphosed gabbro and related rocks
- Metamorphosed sedimentary rocks
- Banded gneisses
- — — Outer Hebrides Fault

3. Road to Huisinis
TARANSAY
Tarbert
SHIANT ISLES
4. Luskentyre beach
2. Golden Road
5. Roineabhal
SOUTH HARRIS
Rodel

NORTH UIST

BENBECULA

7. Machair
SOUTH UIST

ERISKAY
8. Traigh Mhor
BARRA

BERNERAY

| 0 | 10 | 20 | 30 Miles |

| 0 | 10 | 20 | 30 | 40 | 50 Kilometres |

The Callanish Stones, Lewis.

Many places to visit have already been described in the foregoing text, so this short chapter just highlights some of the best places to see the geology and landscapes of the area at first hand. The OS Landranger sheets that cover the area are 8, 13, 14, 18, 22 and 31.

Further details of many places of natural heritage interest (geological, biological and landscape) can be found on the SNH portal at https://gateway.snh.gov.uk/sitelink.

1. Callanish, Lewis: this impressive monument made from Lewisian gneiss is a 'must-see' for any visitor to the islands. There is a visitor centre with car parking.

2. The Golden Road from Rodel to Aird Mhighe, Harris: this winding single-track road northwards from Rodel gives access to some of the most dramatic views across the landscapes of Harris and beyond.

The Golden Road, from Rodel to Aird Mhighe, Harris.

3. **The road to Huisinis, Harris:** another winding single-track road that's worth travelling! It gets you up close and personal to some of the most amazing scenery of this National Scenic Area.

Above left. St Kilda.

Above right. Machair.

4. **Luskentyre beach:** this place provides that Robinson Crusoe experience! Miles and miles of golden sands await, with few other visitors to dilute the wilderness feeling.

5. **Roineabhal, Harris:** Roineabhal was smoothed by glaciers and is treeless – see page 12. It is accessible at many points from the road that runs northeastwards from Rodel to Tarbert.

6. **St Kilda:** the islands are owned by the National Trust for Scotland. Access is possible, but it's a trip for only the most committed visitor. The inhabitants of thatched houses are long gone, but their presence echoes down the ages.

The arrival of the Glasgow plane service on the beach at Barra.

7. **Machair in South Uist:** this is almost a continuous strip of beach and fringing grasslands down the western coast of the Uists. Access is readily available along a number of minor roads that run to the coast. The Machair Way has been established, giving pedestrian access along the length of the machair.

8. **Traigh Mhor, Barra:** this is the only beach in Britain that provides a runway for a scheduled plane service. It is a popular tourist attraction that draws the crowds to see the lunchtime arrival and departure of the daily service to Glasgow.

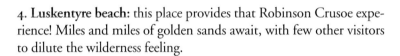

Acknowledgements and picture credits

Thanks are due to Professor Stuart Monro OBE FRSE and Moira McKirdy MBE for their comment and suggestions on the various drafts of this book. I also thank Debs Warner, Mairi Sutherland, Andrew Simmons and Hugh Andrew from Birlinn for their support and direction. Mark Blackadder's book design is up to his usual high standard. Scottish Natural Heritage, in association with the British Geological Survey, published the *Landscape Fashioned by Geology* series that was the precursor to the new *Landscapes in Stone* titles. I thank them both for their permission to use some of the original artwork and photography in this book. Kathryn Goodenough and Jon Merritt wrote the original text for *The Outer Hebrides – A Landscape Fashioned by Geology* that influenced aspects of this book. I dedicate this book to the memory of Professor Janet Watson. She was very generous with her time to me as a young graduate on my first assignment. Her unrivalled expertise in Lewisian geology helped to make my first trip to the Outer Hebrides in 1976 much more useful and productive. Professor Watson was a towering figure in geological science in the 1950s, 60s and 70s. She lacked the pomposity displayed by many of the other senior academics of the time and she was an exceptionally kind and helpful person.

Picture credits

ii–iii Patricia and Angus Macdonald/Aerographica/SNH; 6 Tim Gainey/Alamy Stock Photo; 10 drawn by Jim Lewis; 11 Petr Sommer Photography; 12 Patricia and Angus Macdonald/Aerographica/SNH; 13 (upper) drawn by Robert Nelmes, (lower) drawn by Jim Lewis; 14 Arsgera; 15 Martin Rietzer/Science Photo Library; 16 Dr Ken MacDonald/Science Photo Library; 17 (upper and lower) drawn by Jim Lewis; 19 (upper) Lorne Gill/SNH, (middle) British Geological Survey © NERC, (lower) British Geological Survey © NERC; 20 Alan McKirdy; 21 (upper) Patricia and Angus Macdonald/Aerographica/SNH, (lower) Lorne Gill/SNH; 22 British Geological Survey © NERC; 23 British Geological Survey © NERC; 24 drawn by Jim Lewis; 25 Lorne Gill/SNH; 26 drawn by Jim Lewis; 27 Lorne Gill/SNH; 28 drawn by Jim Lewis; 29 Lorne Gill; 30 Guy Edwards/NHPA; 31 (upper) Peter Lewis/Alamy, (lower) © Newsquest (Herald & Times); 33 John Gordon; 34 Patricia and Angus Macdonald/Aerographica/SNH; 36 (upper) Lorne Gill/SNH; 37: Lorne Gill/SNH; 38 (upper) Ms Mienien, (lower) MichaelY; 39 (upper) Clare Hewitt, (lower) Lorne Gill; 40 Patricia and Angus Macdonald/Aerographica/SNH; 41 corlaffra; 42 (upper) Karel Bartik, (lower) Wolfgang Kruck; 43 (upper) Erni, (lower) Lorne Gill; 44 Alan McKirdy; 45 drawn by Jim Lewis; 46 (lower) markferguson2/Alamy Stock Photo; 47 (upper left) corlaffra, (upper right) snapvision, (lower) Alan McKirdy